SCIENTIFIC LIES

Are Black Holes really Black?

30 Minute Book

A/Prof **Ned Abraham**

MBBS (Hons) MM(Syd) FRACS FRCS(Engl) CSSANZ PhD CAPT (Ret.)

Ned Abraham

Copyright © 2023 Ned Abraham

All rights reserved.
KDP ISBN: 9798385905850
ASIN: B0BXJ48LD8

Books of Controversy & **30 Minute Books**

Books by the same author:

The Clinical Justice System

If you think there is justice in the healthcare system, you better think again! Based on True Events. 174 pages

Simple Answers to the Big Questions

If Science and Religion were contested in Court, They would Both Lose. 523 pages

About the Author

Ned Abraham is a clinical academic who graduated in Medicine & Surgery with Honours (Distinction) then trained as a surgeon in Australia. He studied Evidence and obtained the Degrees of a Master of Medicine and of a Doctor of Philosophy from Sydney University. He is a Fellow of the Colleges of Surgeons of Australasia and of England and a Member of the Colorectal Surgical Society of Australia and New Zealand. He was an A/Professor of Surgery at the University of New South Wales Australia for 17 years and a Captain in the Australian Army Reserve for 10 years. He served both as an officer and as a civilian in Bougainville, Papua New Guinea, and Solomon Islands.

https://en.wikipedia.org/wiki/Ned_Abraham

About this Book

In this 30 Minute Book, we go on a brief journey of some of the lies deliberately told by scientists. In scientific terms, and the same as in a court of law, not telling the truth, the whole truth and nothing but the truth is telling a lie. Some scientists tell lies for personal gains but for many of them, it is a question of "faith". When some scientists "believe" in something, they lie to your face. Most scientists find it very difficult to say, "I don't know", even if that is the truth and they know it. Theorising a concept despite evidence to the contrary and promoting it as "science" is telling a lie. Every scientist must have the guts to stick to factual objectivity regardless of the circumstances and regardless of how illogical or embarrassing the facts might be. A fact is a fact! Scientists should not shape the future. Science should! Science can only do that job properly if "Scientific Lies" were eliminated!

Scientific Lies

The Aim of this Book is to examine the Integrity of Scientists.

This book is dedicated to Simone, Daniel & Aly, Pammie, Corey & Marie...

Please note that the page numbering may differ between the eBook and the paperback versions of this book.

Contents:

About the Author	4
About this Book	4
For Fear of Criticism	7
Holy Science! Unholy Scientists!	9
Richard Dawkins et al	13
Darwin did not *"believe"* in Evolution!	16
Life Ingredients did not exist on Earth!	22
The myth of Micro-evolution	25
Devolution v Evolution	29
We have never been visited by aliens!	31
We cannot colonise space!	33
Discussing "god" is Unscientific!	36
"B-mob Cimota"	41
Humans v *"God(s)"*	43
The Embryos of Haeckel and Richardson	49
The Myth of Natural Selection	51
Neanderthals & Sapiens are the Same!	53
The way we teach Evolution is Dishonest!	54
Humans are *not* the Most "Evolved"!	56
I broke the Laws of Science!	60
The Lies about the Black Hole!	63
Negative Energy and Matter do *not* exist!	70
Let's play the Numbers Game!	73

For Fear of Criticism

Henry Norris Russell was once the director of Princeton University's Observatory. Cecilia Payne, who was born and raised in the UK, immigrated to the USA, and became Russell's PhD student. Cecilia's life overlapped with the lives of cosmology giants such as Percival Lovell, Henrietta Leavitt, Edwin Hubble, Fred Hoyle and, of course, Henry Russell. Cecelia passed away in 1979.

In her autobiography, Cecilia wrote that from the minute she joined Princeton, she learned that Russell's name was synonymous with "American Astronomy". If Russell said "*No*", your paper would never be published!

Cecelia's PhD thesis was on the chemical structure of the stars, the sun in particular. At that time, it was thought that there was a bit of hydrogen and a bit of helium in the sun, possibly 1% to 5%. The other ingredients were thought to be similar to those of Earth.

Cecilia developed a method to quantify the various elements in a star using spectroscopy. When applying her method, she found that hydrogen was by far the most abundant!

Russell did not like it!

Under Russell's instructions, and despite the evidence, instead of setting out to go where no one had ever been, Payne wrote her PhD thesis on why hydrogen could *not* be the main element in the sun! Russell was happy. The paper was published. Payne got her PhD!

Four years later, Russell, and using a different approach, *"discovered"* that over 90% of the sun was in fact hydrogen! The *"discovery"* was even credited to Russell for some years!

The story is self-explanatory! Whether it was control, abuse of power or sexism is a topic for another time. A scientist was forced and eventually agreed to *denying* correct findings at the instructions of her boss! "Knowledge" was put on hold for four years!

Not cool!

Holy Science! Unholy Scientists!

David Attenborough once said, "I think that we've stopped evolving ... we've stopped natural selection ... we started being able to rear 95–99% of our babies who are born. We ... have put a halt to natural selection ... our evolutionary process is now cultural." This opinion is well aligned with Stephen Hawking's theory of "External Evolution" regardless of Hawking's sitting-on-the-fence approach.

Holly Dunsworth and John Hawks were horrified. They're on the attack. They claimed that humans still have a very long way to go in their evolutionary journey! Dunsworth and Hawks argued that some mutations killed certain Humans and others were not compatible with surviving to fertility. Therefore, humans are still being *naturally selected* and are therefore evolving.

Despite agreeing with Attenborough about external evolution, Hawking agreed with the other concept too! Hawking accepted External Evolution while still proposing that humans would still evolve. This is what I mean by the sitting-on-the-fence approach.

It's a mess really!

This may seem like a healthy disagreement between scientists, despite all of them believing in Evolution but in reality, the difference in opinions is a "make or break"!

Why are Hawks and Dunsworth so cranky?

Attenborough made factual observations. He is an evolutionist, but his genius is one of observation. The highest infant mortality rate in the world is in Afghanistan (10%), the lowest in Monaco (less than 0.2%), and the overall worldwide figure is 2.7%. No other living creature on this planet can ever compete with an infant survival rate of 97.3%. Observation suggests that Natural Selection for humans has well and truly stopped, assuming it had started to begin with! No living creature, except for humans themselves, can alter the Natural Selection of humans. Climate change and nuclear holocaust would favour cockroaches over humans. With the exclusion of an asteroid, the only two potentially existential risks to humans are human made! The only *real* risk to humans is humans themselves!

It is not at all hard to see that humans were "naturally selected" from day one! All animals require a significant number of members to avoid extinction, likely to be well over two thousand. If the number drops below a critical level in the wild, a point of no return is reached. If nothing is done, the animal becomes extinct no matter what we do!

In humans' case, DNA evidence shows beyond any doubt that all humans descended from one couple: one male and one female who survived by themselves and propagated the entire species! This is because from the minute they *"appeared"* man and woman were naturally selected because of their off the charts ability to survive!

Why isn't this taught at schools?

If you asked, "Why aren't primates evolving into humans?", Briana Poniner would reply, *"... because they're doing just fine!"*.

I really have no idea what Poniner was talking about when she said primates were doing "*just fine*"! The *"highly evolved"* gorilla is the fourth most endangered animal on Earth! The *"highly evolved"* orangutan is the fifth. Over 300 out of the 633 current primate species are threatened with extinction. Between 25 and 50 primate species are critically endangered!

Primates are most certainly not *"doing just fine"* in the wild, are they?

Why would Hawks, Dunsworth and Poniner tell lies? It is because of their blind *faith* in what they "*believe*" in. Evolution is their religion! Like any fanatic religionist, those scientists have to defend their religion. They have to defend Natural Selection because it's part of Darwinism. They are willing to tell lies to protect their beliefs and spread their religion! Humans were naturally selected from the minute they appeared. It was only a matter of time before they dominated the planet, and they have!

Why isn't this taught at school?

Richard Dawkins et al

Generation of life and evolution criss-cross.

First consider Dawkins' statement:

"It is absolutely safe to say that if you meet somebody who claims not to believe in evolution, that person is ignorant, stupid or insane... or wicked, but I'd rather not consider that."

"Ignorant, stupid, insane, or wicked...". How charming and how desperate!

Next, George Wald wrote that, *"Spontaneous generation* (sic. of life) *was disproved a hundred years ago. But that would lead us to only one other conclusion: that of the supernatural. We cannot accept that on philosophical grounds. So, we choose to believe the impossible: that life arose spontaneously!"*

What Wald said was that, regardless of what you may or may not prove or disprove and beyond any doubt, he would still choose to *"believe the impossible"*. He would not consider an alternative on *"philosophical grounds"*.

How seriously unscientific!

What J Sarfati said was even worse:

"It's OK to deceive students to believe evolution".

A most appalling statement indeed!

We are openly condoning deception! Why are we so blindly sticking to a specific agenda to the point of condoning *"deception"*?

Ironically, what Wald said, and despite his intentions, was that spontaneous generation (of life) was *impossible*! Wald's opinion had nothing to do with factual science and much to do with his religion (Darwinism) and with his own set of "beliefs".

The concept of *"believing in something without evidence"* is highly unscientific!

Over 80,000 generations, the bacterium E coli which has been cultured from a single clone since 1988, has evolved into nothing but E. coli. All the twelve populations grew faster and healthier, but the only thing that E coli evolved into and after more 80,000 generations was E coli itself.

Why isn't this experiment taught at school?

Evolution, if proven, needs to have seriously and effectively navigated five huge mass extinctions. It needs to have recommenced, almost from scratch, at least five times in less than 500 million years. The very short time available was certainly against evolution! Importantly, humans need to have evolved from rats and not from apes!

Why isn't this taught at school?

Darwin did not *"believe"* in Evolution!

Charles Darwin is credited with the theory of Evolution. The words Evolutionism and Darwinism are interchangeable. However, no evolutionary biologist *ever* tells us the truth about Darwin's own opinions! Ever!

You must read Darwin's *"Origin of the Species"* before you can form an opinion.

Darwin just about completely crushed his own theory of Evolution in the last two Chapters of his very own book. Darwin was not dishonest. His current day followers are!

Darwin's extremely harsh and *objective* critique of his own theory of Evolution and in his own book can be summarised as follows:

1. Darwin wrote that we should find innumerable transitional forms, and more so than the actual established species themselves. He stated that their numbers should be *"truly enormous"*, but yet, he wrote, we have found *"none"*. Almost 100% of all current day land vertebrates have been found in fossil format without a single transitional species ever been confirmed!

2. Darwin stated that it was very puzzling that, instead of finding confusion everywhere in nature, all species found in fossil form are *"already well-defined"* and *"perfectly advanced"* for their purpose. No ill-defined, or unadvanced forms have ever been found!

3. Darwin then pointed out the fact that the *sudden* modification of the anatomy of an animal (e.g. tail or wings) by a random mutation cannot suddenly produce another animal with widely different habits. The change in habits would likely need thousands if not millions of generations.

4. Darwin again asserted that we cannot explain how or why natural selection would, on one hand, create something as trivial as the tail of a giraffe which is a mere fly-flapper, and would, on the other hand, produce an organ *"so wonderful as the eye"*.

5. Again, Darwin correctly pointed out that instincts such as parental instincts cannot be acquired through neither natural selection nor sudden random mutations. Reptiles bury their eggs in the sand and walk away. Birds incubate their eggs,

protect them, wait for the chicks to hatch, protect them, feed them and wait for them to become independent. Mutations, random or otherwise, can never do that. If reptiles became birds by a sudden mutation as proposed, they would most certainly burry their eggs in the sand and walk away! Not chick would ever hatch!

6. As for evolution into humans, Darwin wrote, *"Why have not apes acquired the intellectual powers of man? Various causes could be assigned; but as they are conjectural, and their relative probability cannot be weighed, it would be useless to give them."* In other words, if we asked why apes haven't evolved into humans, the reasons would be speculative, imaginary, hypothetical, and can neither be listed nor proven. The reasons are philosophical and are not based on measurable science.

7. Darwin also correctly described how breeding can only happen when members of the same species have intercourse with each other. If two dogs, a Labrador, and a Terrier, have intercourse, they will breed, and the Lab-Terrier will be fertile. Darwin stated that it was *difficult* (it is in fact

impossible) to explain how a newly formed mutated member of a newly evolved species can breed if it were not for the other members of its new species all appearing together, at the same place and at the same time!

Speaking of large numbers of members appearing at the same time, fossil evidence very clearly shows that thousands of members of any new species *did* appear *together* and *suddenly*, i.e. at the same place and at the same time. It is statistically impossible for all of them to have *randomly* mutated in an *identical* manner, let alone at the same time and place on this massive planet!

8. Darwin explained that even if a member of a new species successfully interbred with a member of the last one, their offspring would certainly be sterile (e.g. fertile horse + fertile donkey = sterile mule). The new species would be eliminated! When a member of a new species appears, it will either die a virgin, die childless, or produce sterile offspring which would certainly die childless!

9. Again Darwin wrote that he could not explain how or why Natural Selection, for example dark surroundings, could cause

one animal to produce a powerful eye to overcome the darkness and another to lose their eyesight almost completely!

10. At the end, Darwin added that:

"... *science as yet throws no light on the far higher problem of the essence or origin of life.*" Nearly 200 years later, and the *"far higher"* problem remains totally unsolved by science!

Now we comment: The IUCN Criteria strongly suggest that an animal, e.g. humans, would need a population of at least 2,500 to maintain its species. The Criteria apply to all mammals. However, they clearly contradict the irrefutable DNA evidence that all humans descended from one man and one woman! Propagation was possible for humans simply because they are the most powerful and the most capable creature to have ever existed!

For all other mammals, 2,500 members would be need to propagate or to perverse a species. Thousands of mammals needed to have appeared simultaneously at the same place. This is simply supported by fossil evidence!

The theory of *propagation* of a species other than Humans having supposedly mutated from another, through one male and one female, is virtually impossible!

When you add the fact that 96% of mutations are lethal and the rest extremely harmful, you have an unsolvable dilemma!

Why isn't Darwin's own astonishing criticism of evolution taught at school?

Why isn't the near-universal biological consensus against the theory of spontaneous generation of life taught at school?

Why isn't the fact that propagation requires at least 2,500 members of the same species appearing simultaneously and at the same place taught at school?

Life Ingredients did not exist on Earth!

Regardless of how life started, and unless life came from outer space, you must agree that the ingredients of life needed to *exist!* Soil, water, and/or air! Mustn't you? You'd also expect the elemental composition of soil, water and/or air to be a bit similar to that of living creatures, wouldn't you?

To begin with, Life certainly did come from outer space!

Why? Our carbon-based life would never survive the intense radiation in outer space and would never survive 1,500 degrees centigrade as it enters Earth's atmosphere!

Life formed on Earth!

Earth was formed of the heavier elements, in particular, iron. Earth is a young planet. It is extremely important to realise that everything on Planet Earth is finite! The Planet hardly receives any new matter from outer space via meteorites and asteroids and it loses some gas molecules which escape gravity and are lost into outer space at altitudes of over 150 km. As far as the overall collective elemental content of *everything* on Earth is concerned, and except under exceptional circumstance such as a hydrogen bomb, this is it! Earth is completely finite! There is *literally* nothing new under the sun!

There is a complex life cycle for every element on Earth, whether in isolation or as part of a compound, whether in gaseous, liquid, or solid form, whether living or dead, and whether it is in the atmosphere, in water, in the soil or a combination of all three. The cycle repeats itself!

Let's see!

Earth's centre is 80% iron! Earth's crust is 27% silicon and zero carbon. All ground and underground carbon has come from dead creatures: animals and plants. By weight, plants are 27% carbon, animals 17% carbon, and both groups: almost zero silicon!

Very ironic indeed!

1. Earth's soil which has never seen or been exposed to life has no carbon.
2. Water which has never seen or been exposed to life has no carbon.
3. There is no carbon in Earth's core.
4. There was carbon dioxide in Earth's atmosphere before life appeared 3.5 billion years ago.

Before life began, there was a total of about 5 x 10^{14} (14 zeros) kilograms of elemental carbon available in or on Planet Earth!

If every single atom of carbon on Planet Earth that had ever existed 3.5 billion years ago was directed towards generating life, there would be no carbon dioxide left and we would still be half a million times short of carbon! The most fundamental ingredient of life was all but completely absent!

Finally, the chemical formulas for a transformation from physical to organic chemistry do not exist, neither practically nor theoretically! The three astonishing facts are:

1. The distribution and proportions of available elements in soil, water and air do not have any correlation with the elements of a living organism, whether it is a shrub, or a human being,
2. a process for how life could have "happened" does not exist, and
3. carbon is an unsolvable dilemma.

Why aren't these *facts* taught at school?

The myth of Micro-evolution

Stephen Hawking suggested that life formed as a random cosmic event. RNA formed from inorganic matter. RNA became DNA. DNA became living cells! Simples! Then, *"the rate of increase of DNA complexity gradually rose to about one bit a year"*, but only in the last five million years. All of this has been proven to be overwhelmingly wrong, and well before Hawking's time!

The largest genome currently in existence is *the* most ancient! Lungfish genome is 43 billion pairs of DNA. At Hawking's most accelerated rate of complexity of one bit of DNA per year, lungfish would have taken 43 billion years to develop. Despite being one of the most ancient creatures on Earth, lungfish remains alive and well!

Hawking would have understood, and only too well, the Alan Turing Halting Problem. It is impossible to know ahead of time if or when an algorithm could ever become complete. Hawking chose to ignore the fact that nucleic acids (RNA and DNA) are the only structures known to be Turing Complete! A software that can read absolutely any nucleic acid language. You insert insulin DNA in a bacterium, that bacterium will make insulin!

DNA is the only Universal Turing Machine!

Why isn't this taught at school?

Randomness cannot produce the *"extremely fine adjustment"* described by Hawking. Can I assert that the *"extremely fine adjustment"* seen in an iPhone 17 is a random cosmic event? Luck of the draw? I'd be sent to an asylum, wouldn't I?

As well as the fact that a link between inorganic and organic living chemistry does not exist, a virus cannot exist on its own, independent of a living cell. A virus cannot therefore be the precursor to a living cell. The knockout blow to Hawking's theory is that living cells have existed for 3.5 billion years while viruses have only been around for 1.5 billion years! Living cells had existed for two billion years before viruses even appeared.

Why isn't this taught at school?

Eventually, Hawking admits, *"We don't know how DNA molecules first appeared!"*

Why isn't this taught at school?

Neither science nor mythology provide us with a *chemical* explanation of *how* life began. Scientists only pretend!

Why isn't that taught at school? No matter which side you are on, it seems to be a simple matter of *faith*.

However, it's probably a no brainer!

Life was *"made"*. The ingredients which did not exist were *made* available by someone who knew how! The "who" is not the subject of this 30 Minute Book. Life was *"extremely finely adjusted"*. Nothing random! Next, living cells are two billion years older than viruses. They appear *suddenly* in fossil format and only 500 million years after Earth was formed. All animals and plants appeared just as *suddenly*. Viruses did not "micro-evolve" into living cells because the latter predated the former by two billion years!

Five massive extinctions followed. Sixty-six million years ago, all but the tiniest mammals, such as rats were eradicated. If humans evolved from other mammals, they needed to have evolved from rats in less than 66 million years!

Why isn't the fact that it is chemically impossible for life to *"develop"* in 500 million years taught at school?

Why isn't the fact that life can only be obtained from pre-existing life taught at school?

Hawking eventually agreed that *"Maybe the probability of life spontaneously appearing is so low that Earth is the only planet in the galaxy (or in the observable universe) on which it happened"*.

The scientific stand for almost 200 years has been that life did not spontaneously generate itself!

Why isn't this taught at school?

Devolution v Evolution

The theory of Entropy (chaos) is supported by Hawking and many others. As explained in the last chapter, living cells are two billion years older than viruses! A plasmid is a cell with DNA but without a nucleus. A plasmid is halfway between a living cell and a virus. Somewhere between 3.5 billion and 1.5 million years ago, some living cellular nuclear matter were subjected to entropy! It deteriorated into plasmids. Plasmids can survive independently for a bit longer than viruses.

About 1.5 billion years ago, plasmids deteriorated further into DNA viruses which are completely dependent on living cells! They can only survive on their own for a very short period of time and can only multiply and propagate inside a living cell by overtaking then killing that cell, or at a minimum damage it beyond repair! Viruses then hop from one living cell to another. Millions of random mutations later, they became virulent and started causing devastating pandemics. Some DNA viruses deteriorated further into RNA viruses. There is an element of randomness. First there was order. Chaos followed. Chaos is random!

It would be heretical to propose that living cells micro-evolved from viruses as living cells appeared first, viruses followed two billion years later, and viruses cannot survive or multiply outside of a living cell anyway.

The truth is that some living cells *de-volved* into plasmids. Some plasmids *de-evolved* into DNA viruses. And finally, some DNA viruses *de-evolved* into RNA viruses.

Micro-devolution is proven.

Micro-evolution is a farce!

Why aren't we teaching this at school?

We have never been visited by aliens!

The US Congress investigated UFOs in 1972. No conclusions were reached, and the matter was put on the shelf for 50 years. When the matter was reopened in 2021, pilots and astronauts reported several accounts of bright flying objects caught on video that did not resemble any known earthly aircraft. The USAF personnel reported that the UFOs were moving in a deliberate manner, at extremely high speeds, suddenly stopping and starting, and that they all appeared harmless. There have never been any reports of any incidents of any of those flying objects hanging around for a chat, communicating with USAF personnel, or taking offensive positions against a USAF aircraft. The final report to the US Congress was that there was no evidence that UFOs carried aliens. UFOs were deemed harmless. It seems clear that UFOs are a supernatural phenomenon that science *cannot* explain, possibly refuses to explain in case the outcome embarrassed some scientists. The fact that UFOs are "harmless" does not mean that they're not alive! Their actions indicate deliberate decisions made and deliberate actions taken.

Are they a non-earthly form of life denied by earthly humans? Probably!

As for extra-terrestrial life, there is no reason as to why the *"grand design"* as described by Hawking couldn't have been grandly repeated elsewhere in the cosmos! It is arguable that if an advanced life existed elsewhere, if would have contacted us by now!

We have no way of telling any time soon if life does or does not exist elsewhere. If whoever set up carbon-based life on Earth has done the same thing elsewhere in the Universe, that other life would be as incapable of travelling across galaxies as we are.

We also have no way of telling if another form of life which does not require carbon could have been made elsewhere in the cosmos!

None of those extra-terrestrial scenarios explain the documented UFO sightings investigated by the US Congress. This leaves us with only one conclusion: UFOs are an alternative form of life, here on Earth, that we know nothing about.

Why aren't we teaching these facts at school?

We cannot colonise space!

The nearest planet that is placed in a circumstellar habitable zone of a star for carbon-based life is *probably* Proxima d which orbits the star Proxima Centauri. Proxima Centauri is 4.2 light-years from the Solar System in the Milky Way.

The unmanned Voyager 2 was launched in 1977. It has travelled at an average speed of 3 Astronomical Units per year. An AU is the distance between Earth and the sun which is 12 light minutes. Voyager has travelled about 119 AU before it officially left the solar system 41 years after it left Planet Earth!

One lightyear is over 63,000 AU.

A human spaceship's best scenario would be to *possibly* reach the nearest *possibly* habitable planet in a *possibly* circumstellar habitable zone in *possibly* 89 thousand years.

For this to work, it could be *theoretically* possible to suggest sending a family out there. They would have to generate their own oxygen, food, water, grow up, breed, die, bury their dead, maintain their species for 89 thousand years before they reach Proxima d. I suppose they could carry frozen embryos with them! It would take that family longer to reach

Proxima d than the time that has passed since humans left Africa!

If they got there, only to find that we were wrong, they could either die there or try to turn around and spend another 89 thousand years to return to Earth doing the same things during their return journey!

Quite bizarre!

Alternatively, and to reach Proxima d, humans would have to travel at the speed of light for 4.2 years which is clearly impossible because at the speed of light, we become light!

Hawking's suggestion was to use genetic engineering to make Humans live for 100,000 years. Even if the impossible became possible, who would want to travel in a spaceship for 89,000 years for the sake of making it to Proxima d in case it was habitable? I suppose they could theoretically live there for last 10,000 years or their life!

Moderately insane!

It has been argued that exploring the bottom of the ocean at 10 km below sea level is much harder than exploring outer space. Several parts of Planet Earth remain largely unexplored. The argument that one day there may be way too many humans to live on Earth and that we have to colonise space is academic because we cannot colonise space! Humans will continue to multiply to their capacity. When capacity is reached, they will be slowed down by the finite material of our finite planet.

Exploring space is cool, but why aren't we teaching our kids at school that colonising space is a myth and that they should focus on looking after our planet?

Our planet is all we've got!

Discussing "god" is Unscientific!

I care about honesty and facts far more than I care about whether god exists or not.

The *plausibility* of the existence of a god or lack of it thereof, is purely philosophical. The scientific question is whether she exists or not. It is not whether she *can* or *cannot* exist! Bertrand Russell summed up the difference when he wrote that, *"Science is what we know. Philosophy is what we do not know.*

Simply put, the assertion by Hawking that god *cannot* exist can only be based on *strong faith, rigid philosophy* or both but certainly not on *science*. Whether she exists or not, the "assumption" that god *cannot* exist, as put forward by Hawking and others, is highly presumptive and therefore highly unscientific.

Emotions and feelings are more powerful than thoughts and ideas, even for scientists! Hawking's *strong philosophical faith* in the lack of existence of a god overruled his scientific genius and allowed him to make the highly unscientific statement that *"God cannot exist"*.

It is impossible to prove the lack of existence of anything! We can never search every single square metre of Planet Earth as well as that of the entire universe to prove the *lack of existence* of anything! This seems convenient for theists with the added benefit of their claim that God cannot be seen or felt with our earthly eyes or hands and cannot be detected using a Geiger Counter, or the like.

It is therefore illogical and highly unjust to demand that an atheist proves the lack of existence of a god beyond reasonable doubt using science. For the sake of completeness, one must explore the odds of *disproving the lack of existence!*

Hawking's use of the phrase, *"the grand design of the universe"* which he *admired,* dictates that we determine who the grand designer was. If the Universe was both the cause and the effect as Hawking claimed, then the universe created itself. The universe is *the grand designer* with the *will* to make decisions and the *power* to act on those decisions. The universe would be god! If the universe is god, then there is a god after all! Her name is "Universe"!

Hawking's alternative theory was that the Laws of Science could be god. If the Laws of Science are god, then there is a god after all. Her name is "Laws of Science"!

Hawking asserted that even god herself, if she existed, *could not* break the Laws of Science.

Fair enough!

Almost in the same breath, Hawking then asserted that science *did break* when the universe was being formed! Allegedly as the universe was travelling through its birth canal, science was suspended. General Relativity, the most important Scientific Law of the twentieth century, simply *broke*! Hawking and others have also asserted that science breaks at all *other* points of singularity every time a Black Hole is born! Space-time allegedly ceases to exist!

Hawking's genius mathematically showed what happens inside and outside of a Black Hole. A most amazing mathematical formula! However, *breaking science?* That's a real cop out, isn't it?

For the universe to "create itself" without any input from anything or anyone else, science has to break! Simples!

Hold on a second!

So, god, if she existed, would not be able to break the Laws of Science but Hawking can?

Did science really break? Who broke it? Why?

If the TV "broke", it's because someone or something broke it! A child! A power surge! Even if it was simple "wear and tear", someone or something broke the TV!

Did the TV break? Maybe it didn't! Maybe science did not break after all! Maybe there is another explanation that is (more) scientific as opposed to *"science broke"!*

When we consider the origin of life, life did not come from outer space. There is no scientifically plausible process, whether physical or chemical, experimental, or theoretical, that could show that life began as a random cosmic event. The main ingredient, carbon, did not exist (well, we were half a million times short of carbon)! The composition of life is fundamentally different to that of Earth. Finally, the sequence of events is a complete reversal of the proposed theory of microevolution!

Even if the ingredients did exist (which they did not), the likelihood of the ingredients congregating together at the same place and at the same time is prohibitively minute! The odds are one chance in several google googolplexes.

Hawking then stated that *"reality might be known to God, but the quantum nature of light would prevent us seeing it"*. The proposal is that we are uncertain about absolutely everything. The assumption is that we would never be certain about anything at all! Relativism, by definition, dictates that Relativism itself is relative! Relativists cannot, by their own definition, be certain about their own theory of Relativism!

There is/was a designer!

Her name and her other features are not the subject of this book.

"B-mob Cimota"

Contrary to popular belief, the Big Bang was a completely silent and extremely cold process which happened in total darkness! It perfectly fits in with the transformation of energy into matter following the *scientific law* of General Relativity which is associated with a massive loss of energy. We now know for a fact that the Higgs boson's subatomic energy is what gives matter its weight.

On the other hand, the transformation of matter into energy is associated with a massive release of energy! Atomic Bomb!

Matter had to have come from energy. Considering the size of the cosmos (what we can observe of it anyway), the energy needed to be infinite! Furthermore, not all the infinite energy could have been transformed into matter because we know that *more* energy was required for the stellar blast to commence a billion years later!

Big Bang is the reverse of an Atomic Bomb!

I call it *"B-mob Cimota"*!

Matter, space and time are intertwined! Before matter, neither space nor time existed. Since matter has been here, space-time has *always* been here. If and when matter is completely obliterated back into energy, space-time will cease to exist!

We do not know who provided the infinite energy for the Big Bang or who provided the massive influx of further energy for the stellar blast to take place a billion years later! What we do know is that matter needed to have come from energy as per General Relativity and the Higgs boson, and that the stellar blast required a major influx of further energy.

This the only *scientific* way to explain it!

Why isn't the scientific laws of General Relativity and of the Higgs boson taught at school in the correct manner?

Humans v "G*od(s)*"

Where do Humans fit in the animal kingdom?

In terms of external appearance, Humans look most like the chimpanzees. However, and even according to Evolution, *humans did not descend from monkeys or any other current day primate*! The theory is that both humans and monkeys have a common ancestor which lived about 7 million years ago.

Why isn't this theory spelt out at our schools?

Why are we led to believe that humans evolved from apes when even evolution itself does not suggest that they did?

The theory is that humans and apes are cousins! They have a common ancestor. Who the common ancestor was is anyone's guess! No one knows! That common ancestor has never been found in fossil form. Lucy and Ida were proposed as candidates at some stage! They both had their 15 minutes of fame. Ida was a lemur in icy Germany with a massive bushy tail that was longer than her own body of 24 centimetres. She did not fulfill the *any* of the criteria other than having lower incisors. Lucy was a monkey in Africa! So who was it?

To make a distinction between Humans and other species, one should consider external appearances, DNA genomes, and collective behavioural abilities. Consider the following:

If a scientist from outer space examined sets of data from Earth, on paper, without meeting any of the mammals or learning anything about their behaviour, she would observe the following:

The genomic difference between Humans and Neanderthals is 0.3%.

The genomic difference between humans and bonobo is 0.4%.

The range of genomic difference between current day humans themselves is 0.5%.

Humans and gorillas: about 2% and so on…

Current human genome is 2% Neanderthal.

The variability in DNA between humans is wider than the difference between the average human and the average Neanderthal. This means that Neanderthals were more similar to humans than humans are to themselves. Humans carry diluted Neanderthal DNA!

The alien scientist would conclude that humans and Neanderthals are one species!

She would also opine that bonobo must be *nearly* equal to both humans and Neanderthals. She would conclude that all three must look and behave in about 99% similar ways.

However, when the alien scientist examines behaviour, intelligence, and dexterity, she will have a seriously different story to tell!

99.99% of average humans would score higher than 55 IQ points. The most intelligent ape in history would score about 25 points but so would the octopus, the elephant, and the dolphin! The crow would score higher than all those animals with a massive gulf between the crow and humans.

The gap between humans and the absolutely most intelligent animal is more than four standard deviations from the average. In statistical terms, this means that we are dealing with two completely different independent sets of data. The dumbest human to have ever lived is incredibly smarter than the smartest animal to have ever lived!

It is not a spectrum. It is either that Humans are in a fundamentally different category of mammals, that our understanding of DNA genomes is fundamentally wrong, or both!

Where is the Bonobo New York City that is 0.4% less *"evolved"* than its human counterpart?

The concept of Humans being just another mammal may have been intended to be "humbling", but it is not! As opposed to *all* animals, humans do *all* of this: They think independently, imagine, calculate, assess, make decisions, take calculated deliberate actions, exercise their immense powers, lie, and feel guilt (except if you're a psychopath of course)!

The number of variations in DNA sequencing that every single human, whether male or female can produce, while embarking on the process of copulation for the preservation of the species, is an estimated ten to the power of 706 (ten with 706 zeros beside it).

The eighty billion humans have walked this Earth have all been different to one another and everyone one of them has had an almost unlimited number of potential variabilities in their offspring. DNA is the only Turing Complete software and the only *truly*

Universal Turing Machine to have ever existed! The number of different pieces of information, structures, and combinations of bits of information and structures that DNA can manage, read, and process is infinite.

Finally, humans have undergone less than 0.01% genomic change in over 200 thousand years. No evolution has taken place in 0.2 million years!

Why isn't this taught at school?

Hawking argued that we might have to wait! If evolution was the correct theory (which I opine that it is not), the change from unicellular to multicellular organisms took two and a half billion years to allegedly occur. Even if evolution was true, the sun will die and so will the rest of us before humans evolve into something else!

Hawking would likely say, "Would've!". "Would've" just doesn't cut it!

Why isn't this taught at school?

Crocodiles survived the asteroid of the Gulf of Mexico 66 million years ago. In 145 million years, crocodile DNA has not evolved into anything else!

Did Evolution *forget* to modify crocodiles?

Why isn't this taught at school?

Swine heart valves were successfully transplanted to human hearts to correct abnormalities. The anatomy and physiology of Humans are extremely similar to those of other mammals. This is used by some to prove evolution and by others to prove a common designer.

The answer is academic because Evolution is rubbish anyway and as per Darwin himself!

The Embryos of Haeckel and Richardson

In 1847, Ernest Haeckel, and in support of evolution, drew a famous diagram of embryos of different animals, at the same stage of development. The drawn embryos looked remarkably similar. Haeckel considered the similarities to constitute irrefutable evidence for evolution.

A hundred and fifty years later, and in 1997, Mike Richardson published photographs of the embryos of those same animals at the very same stage of development which were remarkably different to each other as well as to Haeckel's diagrams!

Richardson concluded that Haeckel was *"a fraud"* and his diagrams, *"fake"*. The contrast between the diagrams and the real photos was astounding.

Ironically, both Haeckel and Richardson considered their findings, fake or otherwise, to constitute irrefutable evidence for evolution. In other words, the remarkable similarities between the embryos in the fake diagrams of 1847 were considered irrefutable evidence for evolution, and the remarkable differences between the embryos in the true photographs in 1997 were also considered irrefutable evidence for evolution!

Why would a respected biologist fake the images to prove evolution? How can two sets of data which completely contradict each other, provide evidence for the same theory?

Highly ironic indeed!

Why isn't this taught at school?

The Myth of Natural Selection

Natural Selection eliminates the vulnerable, or the unsuitable. The fittest survives but not necessarily the strongest or the most "evolved". The Mexico asteroid 66 million years ago preserved the tiniest mammals but eradicated the dinosaurs. Natural Selection *produces* nothing. Mutations are exceedingly rare. 96% of mutations are lethal. The other 4% are extremely harmful. It seems extremely unlikely that a mutation could be not only survivable but also highly beneficial to produce a more advanced animal.

The theory that we haven't found transitional designs because they were incompatible with survival is self-defeating! If they could not survive, how could they have propagated their species?

The current land vertebrate fossil record is almost 100% complete. Confirmed intermediate species have never been found in fossil form. As per Darwin, all animals including Humans *appeared suddenly* and when they appeared, they were already in a *well-defined and perfectly advanced order*.

The times of origin of land vertebrates continue to be pushed closer and closer to each other. The times of origin of land vertebrates continue to be pushed backwards and towards the Cambrian Period. As well as implausible, the odds of one ape embryo randomly mutating into a Human are practically zero. The odds of that one Human, regardless of its gender, breeding with an ape and producing fertile offspring are non-existent. The odds of two identical but random mutations occurring simultaneously are prohibitively minute. The odds of those two identical mutations producing the two genders, male and female are too small to calculate. The odds of all those random but identical events occurring at or near the same place and within a few decades of each other, in only a few-million-year history are too remote to exist.

The theory of random mutations followed by Natural Selection and new species is rubbish!

Neanderthals & Sapiens are the Same!

We now *know* that all male Humans who have ever lived on Planet Earth share the same ancestral father by studying the Y-chromosomal DNA. All females who have ever lived share the same ancestral mother by studying mitochondrial DNA.

The original Humans couple had in its genome all the original features of all the human races. As explained earlier, the number of variations in DNA sequencing that every single Humans, regardless of gender, can produce is practically incalculable. A men makes trillions of different sperms in his lifetime. A woman starts with about 6 million different eggs and ends with none by the time she reaches menopause.

The differences in DNA genome amongst all humans are up to 0.5%. The difference between the average human and the average Neanderthal is 0.3%. This means that the average current day human is more similar to a Neanderthal than some other current day humans. The current day human is also 2% Neanderthal. Finally, current day humans bred freely with neanderthal. This proves that Humans and neanderthal are the same species! Why isn't this taught at school?

The way we teach Evolution is Dishonest!

In a very "informative" illustrated book by L B Halstead, published in Australia, titled "The Evolution of Mammals", targeted at teenagers with an interest in evolutionary biology, Halstead asserts that evolution of mammals over the course of about 200 million years has been *"proven"* except for a *"gap"* of 195 million years where fossil evidence is lacking!

Hold on a second!

If 97.5% of the linkage is missing, how could Halstead logically assert that this story of evolution has been proven?

In an exhaustive diagram, Halstead suggested the existence of one possible but very weak and highly interrupted link between marsupials and their ancestors.

Halstead then unequivocally documented the complete lack of linkage between any of the 29 placentals and any of their proposed ancestors!

In summary, out of a 200-million-year history of this proposed particular phase of Evolution, the gap with a complete lack of evidence is *"only"* for 97.5%, and 29 out 29 necessary links for placentals are non-existent.

Why isn't this taught at school?

Are we deceiving our children?

Yes, we are!

Humans are *not* the Most "Evolved"!

In terms of intelligence and fine-motor skills, Humans are not part of a hierarchy because humans are a very extreme "outlier" which is off the charts! All so called "intelligent" animals fall well outside the 4th Standard Deviation of human intelligence. This means that the most intelligent animal on the planet is far less intelligent than the least intelligent human on the planet. In statistical terms, the two sets of data are not a "continuum". We are dealing with two completely different sets of data! Not a spectrum!

The "hierarchy" proposed by humans, in the animal kingdom, is based on who eats whom! It's based on the premise of predator and prey. Humans decided that tigers are more advanced than gazelles because tigers eat gazelles. Humans have the power to eat anything, including tigers!

Very dark indeed!

However, a cheetah can easily chase, catch, and eat a chimp (even a human for all I care), yet we claim that chimps are more evolved than a cheetah simply because they look more like us than a cheetah does!

Now consider this:

The dolphin, elephant and octopus are all as smart as the chimp!

Despite a massive gap between it and humans, the crow is the second smartest creature on Earth after humans. The crow *should be* considered more evolved than the chimp, shouldn't it?

Now, eagle's eye is extremely more powerful than the human eye. So, who is more evolved? Kangaroos can hang onto their joeys and not give birth to them when they're due if there is a drought! Can humans do that? Who is more evolved? Humans cannot fly without machinery! Who is more evolved then, humans or birds? A flea can jump over 4 metres. If a human had the same power in their legs, size for size, as a flea, she would easily jump over the Empire State Building. So, who is more evolved? The Central American lizard can walk on water. So, who is more evolved? The elephant's ear is far more sensitive than any human ear. So, who is more evolved?

Although dung beetles become completely lost in a moonless night, when the moon is bright, dung beetles move in a perfectly straight line guided by "polarisation" using moonlight. How can humans compete with that? Who is more evolved? The cheetah can reach and sustain a speed of 110 kph. The fastest man in history, Usain Bolt, ran at a speed of 36 kph for 10 seconds then stopped to catch his breath. So, who is more evolved?

The bloodhound's nose is about one million times more sensitive than the human nose. So, who is more evolved, humans or dogs? When a bee finds nectar, she returns to the colony, goes straight to the centre of the group, and dances! Without "talking", the orientation of her dance tells the other bees whether the nectar is in the direction of the sun or away from it. The speed of flapping her wings tells them whether the nectar is far or nearby! Can a human do that without "talking" or "writing"?

Bats emit ultrasonic waves and use their echoed vibrations to locate and hunt tiny insects using echolocation. Astounding! Can human perception compete with that? So, who's more evolved?

The reality is that, as Charles Darwin correctly pointed out, all animals are already *"perfectly designed and perfectly suited for their job*! Even Darwin accepted the perfect design concept! When they "appeared" in fossil form, all animals were *"already in a perfectly complete order"*. Same as everyone else, when humans appeared in fossil form, they were *"already in a perfectly complete order"*.

Charles Darwin was honest. His current day followers are not! *Faith* is what causes those who *"believe in evolution"* (Richard Dawkins' own words, not mine) to interpret the evidence the way they do. Humans live a Big Scientific Lie that they are the most *"evolved"* creatures on Planet Earth. It is a Big Lie because the concept of *"Evolution"* is completely dependent on how you define it! More often than not, that definition is made-to-measure! It's made to fit whatever argument is topical!

Charles Darwin tore down his own theory of Evolution! Darwinian Evolution is rubbish!

Why isn't Darwin's own book studied at school?

I broke the Laws of Science!

Laplace, despite being a Christian, and during the reign of Napoleon Bonaparte, indirectly opined that even God, if she existed, could not break the Laws of Science, an opinion strongly shared by Stephen Hawking. Hawking consistently argued that the position of every scientist must be that *"a scientific law is not a scientific law except if it holds regardless of any external or supernatural intervention"*.

When asked about god, Hawking asserted that *the way* the universe began was not chosen by a god but determined by the Laws of Science. He agreed with Laplace that the Laws of Science cannot possibly be broken and even wondered if the Laws of Science could be called 'God'.

Furthermore, when Hawking researched the origins of the universe, he argued that the most fundamental objection to the universe having been created is that creation would be a place where *"science broke down"*.

For Hawking, this was irrefutable evidence that the universe was not created.

Hawking then used the "theory" of anti- (negative) matter, to suggest that the universe came from "Nothing" because "Nothing" produced two equal amounts of "positive" and "negative", i.e. matter and anti-matter.

In other words he suggested that because 1 + (-1) = Zero, then you must be able to get 1 + (-1) from zero! Total rubbish indeed!

Hawking then confidently asserted that at singularity (the point origin of the universe), Einstein's Theory of General Relativity *"broke"*! Astrophysicists in Hawking's camp also assert that: *"The rules of physics do not exist in singularity."*

It gets better! Apparently, the Laws of Science *do* break again and again every time a star dies, and every time a Black Hole is formed! Apparently, time and space also cease to exist!

Hold on a second! So, science *does* break after all? It's a dumb play on words really! So, god, if she existed, could not break the Laws of Science by Stephen Hawking can!

Hawking hinted that when a Black Hole is formed, matter and energy cease to *"change"*. However, that's not true either! Some of that matter and some of that energy escape the Black Hole through its Event Horizon!

The space occupied by a Black Hole most certainly does not cease to exist. How would a Black Hole move around in space and grow in size over time by engulfing other celestial bodies if it wasn't subject to time and space?

A Black Hole is a balance of what it engulfs minus what escapes it! As Hawking once said, "You'll eventually get out!". Maybe, science does not break after all! It is true that in the physical world, science *is* unbreakable except if you were Stephen Hawking, I suppose!

If the Laws of Science did not exist when the universe began, try to answer the following questions: *Where did Science come from? When did it begin? How? Who made it? What rules guided the universe before science existed? How and when did that other set of rules come into existence? When did it cease to exist? Why were those rules replaced by the Laws of Science? What was wrong with them? How did they cease to exist? Who decided? Did something in the universal embryo trigger science? If so, what was it?*

The Lies about the Black Hole!

The universe originated from a massive Black Hole... Did it really?

A Black Hole is an extremely heavy and dense celestial body, a star that had used most but not all of its hydrogen and collapsed under its own gravity. A Black Hole is so heavy, its escape speed is faster than the speed of light. This means that light photons despite being weightless and travelling at a speed of 300,000 km per second, are too slow to escape the gravity of the centre of a Black Hole in total compliance with the Laws of Science. To clarify, Earth's escape velocity is about 11 km/second. The escape velocity of any Black Hole in faster than the speed of light!

However, the event horizon which is the periphery of a Black Hole, or the transition between the dense interior and the empty cosmic space, is a sphere of a shiny red circle surrounding the Black Hole. As the gravity is somewhat less intense, some light escapes the Black Hole emitting Hawking's Radiation, again well and truly in compliance with the Laws of Science.

A Black Hole is a *deteriorated* star! An end product! Hawking's radiations are so minute they are totally unable to create a star or a planet! Although the event horizon is a sphere, it is seen as a circle because the core or centre of the Black Hole casts a shadow on its own shiny sphere which makes the sphere look like a circle. A Black Hole also casts a shadow on other celestial bodies behind it, thus confirming that it's not empty.

Hawking theorised that when a Black Hole stopped spinning, it continued to emit particles of matter and energy. Hawking showed that there was enormous randomness on the surface of a Black Hole (the Event Horizon) in the way it emitted those particles.

This is actually correct (for a change)!

Following the Laws of Science, gravity dictates that the heavier the Black Hole, the lower the energy (temperature and light) and the lesser the amount of matter which manage to escape it. Black Hole the size of the sun emits a fraction of a degree above Absolute Zero. Absolute Zero is minus 273.15 °C. The bigger the Black Hole, the bigger its surface area. Although radiation out of it is of a low intensity, the larger surface area probably makes up for the intensity.

The bottom line is that the bigger the Black Hole, the less likely it is to release anything at all, let alone a star or a planet!

Hawking also suggested that when a Black Hole is formed, space-time cease to exist!

Again this is highly incorrect!

Matter is lost inside Black Holes to us! We cannot see it. However, for space-time to cease to exist, a Black Hole must not change, move, or interact with any external factors. It would there but not there! It would be eternal with no beginning and no end!

Although a substantial force would be needed to move a Black Hole, Black Holes do move in space! Black Holes move within and with their galaxies. At different times, they are at different places. They're subject to time and space. Space matter falls into a Black Hole making it larger and larger over the course of time! This can be in the form of stars, space dust, planets, or smaller Black Holes.

The information is dismantled and stored inside the Black Hole except for Hawking's Radiation. It might take forever for all the contents of a Black Hole to escape its gravity, but, at least in theory, they eventually will!

Forever is an awfully long time but it is still "time"! When Joe Blogs dies, he *never* comes back. This does not mean that time and space have ceased to exist.

Matter could be transformed into energy in several ways and into other matter by nuclear fusion, but it never ceases to exist. Matter engulfed by a Black Hole is dismantled but not eliminated. The information is scrambled but not lost. Matter and energy are interchangeable but neither of them is ever lost! Hawking described this interpretation by using the example of burning an entire encyclopedia but keeping the smoke and the ashes. An alternative example would be the one where you completely melt a fancy custom-made twenty carat gold ring. You permanently lose the ring, but you keep the gold and the alloy in melted form inside the Black Hole.

The information inside a Black Hole is retrievable! This may happen extremely slowly through Hawking's Radiation, but the information is retrievable. Scrambled beyond recognition but retrievable!

Matter and energy are never lost!

Time and space most certainly do not cease to exist when a Black Hole is formed. The suggestion that a Black Hole is at a point where space-time ceases to exist, is wrong. On the other hand, no Black Hole has ever been observed to turn back into a star and to form planets, and there is no theoretically feasible mechanism for this to happen. What is theoretically feasible is that Black Holes could continue to merge into each other and engulf other space objects. The net amount of matter and energy which is the balance of what's engulfed and what's emitted, determine the future of the Black Hole!

Some Black Holes will lose more than what they gain and end up disintegrating. Others will gain more than what they lose and continue to grow larger and mightier!

At the end, the universe will either become one extremely massive Black Hole or return to a sea of matter and energy. In either case, the universe as we currently know it will end!

The universe may also eventually run out of hydrogen fuel and freeze over. How long? Longer than anyone could possibly imagine!

In summary, the formation of a Black Hole is not singularity. A Black Hole is, in a way, an end product and cannot plausibly be the origin of any well-defined matter, let alone the universe. The Laws of Science do not break when a Black Hole is formed.

Can we teach this at school?

Are Black Holes really black?

As explained above, a Black Hole is a dead and burnt-out star that has collapsed under its own immense gravity. Light is too slow to escape the centre of a Black Hole but it's fast enough to escape its event horizon! A Black Hole casts a shadow on its own Event Horizon. The Event Horizon is a bright sphere and not a circle!

Now, the more advanced a space telescope is, the more capable it is of visualising the Event Horizon. Over the course of the last 30 years or so, and with advancing technology, the Event Horizons of the same Black Holes have been observed more and more clearly. The more sensitive the telescope, the wider and brighter the red rings around the Black Holes.

When the technology is advanced enough, we will be able to see an Event Horizon in its entirety for what it is; a sphere! The centre of a Black Hole will no longer be able to cast a shadow on its Event Horizon in front of a powerful telescope!

Black Holes will no longer be seen as being black!

This is because Black Holes are not black!

Negative Energy and Matter do *not* exist!

Energy and matter are the same thing! They cannot be separated. The atomic weight of hydrogen is one. The atomic weight of iron is 56 which means that the weight of a certain volume of iron should be 56 times heavier than that of a similar volume of hydrogen under the same temperature and pressure. However, and under the same conditions, one cubic meter of iron, instead of being 56 times heavier than hydrogen, it is 96,000 times heavier that hydrogen! The difference is in the energy of subatomic Higgs Boson. Energy and weight are inseparable.

Negative Energy in space has never been detected. It cannot be measured. It does not exist. It is *assumed* to exist to balance out the original source of positive energy in the minds of some astrophysicists like Hawking, to make the equation "fit", and to explain the expanding universe at its periphery.

However, it is an illogical concept because normal positive "energy" in the form of "gravity" in the unobservable part of the universe is not negative energy. It's just gravitational energy which attracts the peripheries of the universe and causes them to continue to expand outwards. We just cannot see the matter which is the source of the extra-universal gravitational pull.

Black (Dark) Matter should not be confused with antimatter (negative matter). The latter cannot be measured or detected because it does not exist. Again, it is *assumed* to exist to balance out matter and its source in the minds of astrophysicists like Hawking, to make the equation fit and to preclude an intervention.

On the other hand, Black (Dark) Matter is simply matter that we cannot see! Regardless of its temperature, and even though we cannot see it, Dark (black) Matter has a gravitational pull and that was how we discovered its existence. According to Rubin, *"the ratio of dark-to-light matter is about a factor of ten. That's probably a good number for the ratio of our ignorance-to-knowledge"*.

The Universe has been calculated to comprise of 5% visible matter (Light Matter) with the rest being invisible (Dark) matter and empty space.

The fact that the largest structure in the universe (Hercules-Corona Borealis Great Wall) was only discovered in 2013 tells us how little we know about the universe!

Some of the most ridiculous claims made in the history of science were made by Stephen Hawking. *"The universe simply popped into existence"*. *"The universe created itself"*. *"The universe required neither assistance nor energy to do so"*. *"Nothing caused the Big Bang; absolutely nothing"*.

What a heap of rubbish!

Let's play the Numbers Game!

Finally, we get to play the numbers game!

Scientifically, in *day-to-day life*, if the odds of Event A happening are less than one chance in 10 to the power of eight (1 with 8 zeros beside it), we can *scientifically* conclude that Event A *will not happen!*

In experimental science, if the odds of Event B happening are less than one in 10 to the power of fifteen (15 zeros), we *scientifically* conclude that Event B *cannot happen!*

In theoretical sciences, e.g. astrophysics, if the odds of Event C happening are less than one chance in ten to the power of fifty (50 zeros), we *scientifically* conclude that Event C *cannot happen!*

As per Stephen Hawking, the minutest change in any of the billions of features of the universe would have made the existence of life on Plant Earth impossible. Hawking pointed out that *any* other arrangement and *any* change no matter how minuscule, would have been *too wrong* for life to exist. Like the multiple numbers in a lottery ticket, and as the number of nonviable alternatives is in the googolplexes, the odds of *all* arrangements having come together randomly and by chance are too small to calculate. The odds of the universe having become the way it is by chance are less than one chance in several googol googolplexes. Those odds are prohibitively smaller than the odds of one person winning the jackpot in a million consecutive lotteries without advance knowledge of the winning numbers!

The proposition that the conversion of energy into matter following Einstein's Law of General Relativity and the Higgins Boson happened without a *decision* followed by *action* is *scientifically implausible*. Cause precedes effect. Cause is higher than effect. It is even philosophically illogical to suggest that an effect came to be without a cause.

Claiming that all of those extremely fine arrangements came together randomly would be the same as claiming that Air Force One came together by a series of random accidental events by chance occurrence!

Are we teaching that at school?

Another influx of energy caused the stellar blast to happen a billion years after the Big Bang, thus make the universe almost self-sufficient.

There was no other way for the stellar blast to happen! Can we teach that at school?

The universe was neither formed in an instant, nor was it created in six days 4004 years BCE as suggested by the book of Genesis. The universe was *decisively started* in an instant, most likely 13.7 billion years ago, and it has been expanding following the Laws of Science (physics and chemistry) ever since. It has taken the universe several billion years to get to where and how it is now, and it continues to evolve.

According to Hawking, everything in the cosmos is *finely adjusted*. As explained earlier, the odds of random occurrence of those extremely finely adjusted variables are too small to exist.

Still on the numbers game, we know examine the odds of life appearing the way it did.

Hawking acknowledged that the fact that carbon atoms should exist at all, required an *"extremely fine adjustment of the physical constants"*. As well as the fact that there is no known experimental or theoretical mechanism for the formation of DNA or RNA from inorganic chemistry and guided by the number of bits of information and by the complexity of that information, the odds of an *unknown* process turning simple elements into nucleic acids randomly, are again less than one chance in several googol googolplexes.

Even if the *unknown* process were known, it would have taken at least 43 billion years for the nucleic acids of the most ancient fish, the lungfish, to form.

Again, the odds of amino acids forming randomly on Planet Earth, are less than one in several googolplexes. The odds of simple amino acids randomly turning into simple or complex proteins are non-existent. The odds of the minimum of 100 proteins required by a living cell, congregating together at exactly the same place and exactly the same time in the history of Planet Earth, so they could form the simplest living cell, by chance, is less than one in ten with 2,000 zeros beside it.

Harold Morowitz calculated that the odds of one living cell forming randomly from those one hundred proteins (if they had ever existed together), was less than one chance in ten with 100 billion zeroes beside it.

All amino acids involved in the making and duplicating of DNA, *with no exceptions*, must have left-sided links to fit onto the nucleotides of the DNA which *must* have right-sided links *with no exceptions*. Ralph Muncaster calculated that the odds of this happening as a chance random event are less than one chance in ten with 33,000 zeros beside it!

The statistical probability of the random occurrence of all the cosmic arrangements described, and of the random generation of life may calculated using all the individual odds above by a process of multiplication.

The odds of random occurrence of the setup of the cosmos and of life are one chance in a number which has no name and is too huge to imagine, let alone calculate. Using the *scientific guidelines* listed above, the cosmos and the life it bore were not random events.

Why aren't those numbers taught at school?

Hawking eventually affirmed that we now know how things work *"except for the most extreme situations such as the beginning of the universe"* and that he *"appreciate(s) the grand design of the universe"*.

Finally, something honest was said!

"I don't know" is sometimes the most scientific answer! Many scientists are not true scientists! Sorry, not sorry!

www.ingramcontent.com/pod-product-compliance
Lightning Source LLC
Chambersburg PA
CBHW031534210526
45464CB00019B/1074